BABY ANIMALS

A PORTRAIT OF THE ANIMAL WORLD

Paul Sterry

NEW LINE BOOKS

Fax: (888) 719-7723
e-mail: info@newlinebooks.com

Printed and bound in Singapore

ISBN 1-59764-109-X

Visit us on the web!
www.newlinebooks.com

Author: Paul Sterry

Publisher: Robert Tod
Book Designer: Mark Weinberg
Production Coordinator: Heather Weigel
Senior Editor: Edward Douglas
Project Editor: Cynthia Sternau
Assistant Editor: Linda Greer
DTP Associate: Michael Walther
Typesetting: Command-O, NYC

INTRODUCTION

A mother white-tailed deer keeps a wary look-out while its fawn takes a drink from a river. Although able to run well, the youngster is not yet fully aware of the dangers lurking in the world around it.

For most adult humans, there can be few more precious moments than the birth of their child. With this new life comes renewed hope and the expectation of better things to come. From the point of view of the species as a whole, howev-er, a baby is more than this. As part, albeit a very small part, of a new generation, the new child ensures the continuation of the very species itself through the genes carried in its cells, and the promise of more generations to come.

And so it is with all animals. At the core of the lives of all liv-ing things, the single most important purpose is to reproduce and, if possible, ensure that as many offspring survive for as long as possible—to the point where they themselves can repro-duce. Baby creatures of all species, whether cared for almost until adulthood (as in higher mammals) or abandoned at the egg stage (as in many amphibians), are the key to the survival of their particular species.

Aside from their biological importance, there is no denying that baby animals are cute. With outsized heads and appealing eyes, young mammals in particular can look utterly adorable. In part, this book is intended as a celebration of young animals of all types. Their lives and the lives of their parents will be

described, including preparations for birth and the care, development, feeding, and training that occur in the first weeks or months of a new life. Some of the ways in which the young are nurtured are truly amazing, as are the parental skills and obligations of many lower animals with which parental care is not normally associated. These creatures, from birds, reptiles, and amphibians to fish, spiders, and insects, all have a story to tell in the lives of their babies.

With the exception of the most hard-hearted among us, almost everyone finds baby animals absorbing to watch. This fascination arises, in part, from their appealing features and their vulnerability, but is also due in no small way to their intriguing antics. We can learn a surprising and sometimes alarming amount about ourselves by watching baby animals interacting with their peers, parents, and non-parental adults. In short, they allow us all to play the amateur behavioral scientist. Without necessarily intending or wishing to, we may see aspects of our own family lives reflected in the lives of pets, and among both wild and domesticated animals.

This book tells the story of the lives of baby animals and illustrates the central role that these creatures play in the life histories of all species. It is hoped that the caring role of the parent will also become apparent, as in many species it continues long after the birth itself.

At an age of only four days, these baby blue tits are entirely dependent upon their parents who bring them food, remove their droppings, and keep them warm at night.

Although confident in her own ability to defend herself, this mother cheetah keeps a wary eye open for danger on behalf of her cub.

A NEW LIFE

Almost all female animals modify their lifestyles or patterns of behavior prior to giving birth; in those species where pair-bonding persists beyond conception, both partners may be involved. These changes may be subtle and hard to distinguish until the latter stages of pregnancy, or may be marked and profound from early on. They could simply involve a change in diet, feeding behavior, or choice of habitat, or may involve the construction of a nest or the selection of a safe retreat for the female's confinement. Whatever the changes, however, the intention is the same: to ensure the health and safety of the mother and to secure the safest surroundings for the birth and the first days or weeks of new life. In the latter stages of pregnancy, during the birth itself, and sometimes for a considerable period afterward, the mother is almost as vulnerable to attack as the young to which she is giving, or has given, birth.

Preparation for Birth

Within the mammal group as a whole, each species differs in the way in which it prepares for birth. For some creatures such as antelopes, birth may simply occur where the female finds herself; with herd animals that are constantly on the move, a precise location for birth is difficult to predict. In some antelope species which have roughly synchronized births, however, migration may occur: The whole herd will move to an area of good grazing in time for the births. This will ensure that the females are in the healthiest state possible and will hopefully provide the youngsters with a good diet once weaning has occurred.

Migration prior to the act of giving birth is found in other mammals for different reasons. The humpback whale has been particu-

larly studied and its annual movements are well known. The adults move from their cold-water feeding grounds to warmer, but comparatively food-poor tropical waters to give birth. The reason for the journey, which involves thousands of miles of travel each year, is to ensure that the baby whale can spend the first few weeks of life, when the blubber layer provides inadequate insulation, in optimum water temperatures. The whale calves feed on their mothers' rich milk, and the mothers themselves rely on their stored food reserves.

Among many small mammals, a common strategy in the preparation for birth is to build a nest of some kind. Voles and mice may construct a loose ball of dry grass, lined with hair and placed in a specially excavated burrow or in a hollow tree. Others, such as the diminutive harvest mouse, build intricately woven nests which are lodged among the stems of grass and other vegetation—hopefully out of the way of ground predators.

Large mammals, and particularly carnivores, dig deep and sometimes labyrinthine tunnels in which a secure den or lair is located. Some species, such as foxes and badgers from temperature regions, spend much of the non-active periods of their lives underground and have permanent subterranean homes. Others, including some hyenas and other large predators, excavate special pupping dens; wolves and leopards usually make do with a cave or rock crevice.

An African jacana has built a floating nest for its eggs. Its long toes, which give it the alternative name of lilytrotter, are a bit of a hindrance when trying to settle on the eggs.

Known in Britain as the great northern diver, this common loon is settling down to incubate its eggs. Great care is taken not to get the eggs wet.

Following page: In the wild, baby Nile crocodiles from the same nest tend to hatch at the same time. Just prior to hatching, they emit faint squeaks which are enough to alert the mother to the imminent event.

Nest-Building

The preparations made for birth are never more obvious nor more easily observed than among birds, the most obvious manifestation being nest-building. Unlike almost all mammals, which give birth to live young, birds lay eggs; it is inside these that most of the development of the young bird takes place. It is vital, therefore, that the nest is well concealed in most cases, and secure enough to hold an adult with a full clutch of eggs during incubation, and a full brood of fledgling young in the latter stages.

Nests come in all shapes and sizes. Some are little more than a scrape in the ground, the adult bird's cryptic markings often providing the necessary concealment. At the other extreme are some extremely complex constructions. Many of the most elaborate nests are made by songbirds; these are generally woven structures made from grass stems, twigs, or bark fibers. Birds often conceal the nest within a bush or the branches of a tree for protection from predators. Other extreme examples of nest sites include those of woodpeckers, which excavate holes in trees, or species that use underground burrows. One

of the most unusual types of nest is that of the malleefowl of Australia, which simply buries its egg in a large mound of earth. The male regulates the incubation temperature by adding or removing quantities of soil.

Apart from the few unusual mound-nesting species, all other birds incubate their eggs. This usually involves the female bird but, in some species, both parents may play a role in this process. Special brood patches on the belly ensure that the eggs come into direct contact with the skin; a warm and constant incubation temperature is essential for most, but not all, bird species. Not surprisingly, the time taken from the start of incubation to hatching varies markedly between species. In small songbirds, it may take as little as fourteen days, while in the domestic chicken the incubation time is twenty-one days. In many albatross species, incubation can last for up to eighty days.

Egg-Laying

Among all reptiles, the early stages of development of the young occurs within an egg. Although most species do actually lay eggs, with some, the eggs hatch inside the

A tiny chick is the first of the little grebe brood to hatch. The parent is feeding it tiny scraps of food at this early stage in its life.

Having returned to the beach where she herself was born, a female green turtle is laying her own eggs in a nest dug in the sand.

body of the mother—hence she appears to give birth to live young. Although considerable effort on the part of the female reptile may go into selecting a suitable site for egg-laying, with most species parental responsibility ends there. Such is the case with marine turtles, females of which haul themselves ashore on remote beaches in order to lay their eggs in pits excavated in the sand. Once covered with sand, the female returns to the sea and has nothing more to do with the eggs. Intriguingly, however, her choice of egg-laying site not only affects the survival of her offspring but also their sex: Warmer incubation sites produce males, cooler sites females.

Crocodiles also lay their eggs in the ground beside water, but in the case of most species, the female also guards the site while they are incubating. She will also help the young crocodiles as they hatch, sometimes by digging them out of the ground, and will guard them jealously from predators in the early stages of life in the water.

Similar parental behavior can be seen in some species of fish, although the vast majority which live in shoals produce huge quantities of eggs, and the young that hatch must

Baby green turtles in the same nest all hatch at once, usually at night. They then dig a passageway to the surface of the sand and make their way down the beach to the sea.

With freedom in sight, a baby green turtle has to run the gauntlet of predators such as gulls on its way from its beach nest-site to the open ocean.

Taking its first glimpse of the outside world, a baby Nile crocodile breaks free from its egg. For the first months of life, the mother will keep a careful eye on her offspring.

Considering their relative sizes, it seems remarkable that these baby Nile crocodiles have just hatched from their eggs.

A baby hognose snake emerges from its leathery egg case. Others from the same batch of eggs can be seen emerging in the background.

Aphids are remarkable creatures. They are parthenogenetic and, at birth, baby aphids already have the embryos of the next generation developing inside them. This productivity is needed because numerous predators such as this larval ladybird eat them by the millions.

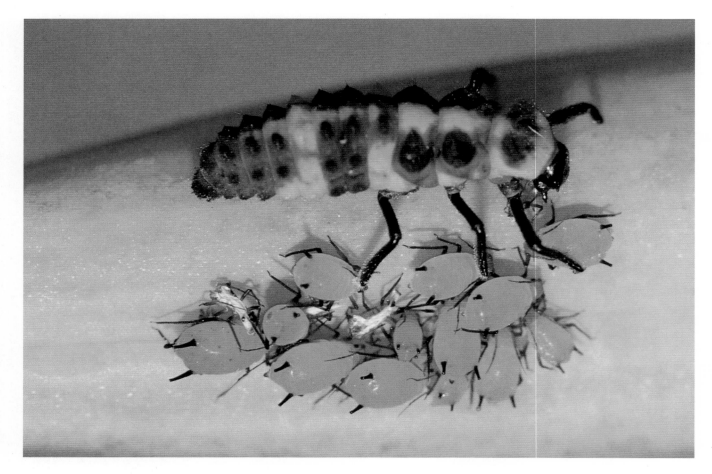

As if their gruesome appearance were not deterrent enough, a mother tailless whip scorpion carries her offspring around for extra protection.

fend for themselves. Male sticklebacks, for example, build a nest in which their mate lays her eggs. Having fertilized the eggs, he then guards them and cares for the young when they hatch.

With amphibians, and with invertebrates too, the provisions made for the young do not often extend further than ensuring that the eggs are laid in a suitable place. With frogs, for example, this might mean choosing a pond which has good algal growth and is not going to dry up in a hurry. With butterflies, as another example, the selection of the correct larval food plant for egg-laying is an obvious way in which the female helps her offspring. Where there are generalizations, there are, naturally, exceptions. These might include the parent bug which guards its eggs, and later its small nymphs, with its life; or the midwife toad, the male of which carries strands of developing spawn wrapped around his body until the tadpoles hatch.

Giving Birth

The majority of mammal species around the world are placental mammals, with most of the development of the young occurring inside the mother's womb and nutrition obtained via the placenta. As a consequence, the young when born are comparatively large and possess all the basic attributes of the adult animal. The process of birth for placental mammals is, not surprisingly, a traumatic time for the mother and one when her life and the lives of her offspring are at great risk.

For antelopes and other open-country animals whose main means of defense is to run, the process of birth is as brief as possible.

A young African spurred tortoise is hatching from the egg in which it has spent the last few weeks developing.

With wildebeest, which are common on the plains of East and Southern Africa, the birth of the single young might take no longer than a few minutes from start to finish. What is more remarkable is the speed with which the baby wildebeest recovers from the shock of being born. Within a few minutes, it has been licked dry by its mother and is making stumbling attempts to stand on all fours. After a period of some 45 minutes, it is able to stand confidently and run; thereafter, it must keep up with its mother even if the herd is panicked into running. Contrast the speed of this with the human baby, which may be a year old before it takes its first unaided steps.

For most other mammals, including those which give birth to more than one offspring on each occasion, the process of giving birth may take considerably longer. Hence the need for secure surroundings such as a den or

lair in which the process can take place in relative safety.

For marine mammals, which still have to breathe air, giving birth presents a few obvious problems. Seals overcome this by leaving the water. In temperate regions, females haul themselves onto remote beaches to pup, while in polar regions, ice floes provide a suitable alternative for many species. Whales, on the other hand, do not have the option to leave the water. Instead, birth takes place in the water, invariably tailfirst, and the mother quickly nudges the newborn to the surface to take its first lungful of air.

Among marsupial mammals, the female appears to treat birth as something of a non-event, sometimes appearing oblivious to the fact that anything at all is going on. The reason for this is that the young is born at an extremely early age and is, not surprisingly,

As soon as they have all hatched, the mother Canada goose leads her brood away from the nest and to water. Unlike songbird chicks, those of wildfowl are active and mobile from the moment they hatch.

tiny. In the gray kangaroo, for example, gestation in the uterus lasts for forty days, after which time the minute, fetus-like young, smaller than a little finger, climbs up the mother's fur and attaches itself to a teat in her pouch. Here it remains for the next two or three months before it takes any interest in the outside world. Contrast this with a similarly sized placental mammal such as a fallow deer, where gestation lasts some 240 days and the young, when born, are fully formed and mobile within a few days.

Arguably the most crucial period for a nesting bird is that when the eggs are hatching; in many ways, this is the avian equivalent of giving birth. For a few hours—or sometimes even days—before hatching, the young birds call from inside the intact egg, giving the parents advance warning of the impending event. Each chick has a special horny plate on the bill, called the egg tooth, with which it breaks open the shell; the egg tooth is lost soon after hatching. The whole process usually takes an hour or so, leaving a rather bedraggled and exhausted youngster at the end.

In some bird groups, such as game birds which nest on the ground, the chicks are soon highly active. Hatching is usually synchronous, so within a few hours, the whole family leaves the vicinity of the nest in search of food; the chicks follow their mother and peck instinctively at anything that could conceivably be eaten.

In the case of other groups, the young birds hatch at a stage when they are incapable of looking after themselves. They must be kept warm by the parents and fed until such time as they can fend for themselves. Among birds where a clutch comprises several eggs, hatching may be synchronous, incubation having started with the laying of the last eggs, or asynchronous, the incubation having begun when the first egg was laid.

Although the development and hatching of reptile eggs shows a lot of similarities with that of birds, the same cannot be said for amphibians. The development of frog spawn is familiar to many people from their school days; cell division and differentiation occurs for all to see in the protective jelly of the spawn. Tadpoles emerge at an early stage, and considerable growth and development, as well as metamorphosis, must occur before a miniature frog is formed.

There are few more endearing sights in nature than of baby goslings. These young Canada geese are enjoying their first few days of life.

Baby screech owls would have little chance of survival without their parents. They are unable to find food until they leave the nest.

Screech owls make excellent parents. This adult bird is incubating her newly-born offspring, which she will defend vigorously against any threat.

Moments after the birth of her foal, a mare instinctively starts to lick the baby clean and dry, and within an hour or so the foal itself tries to take its first steps.

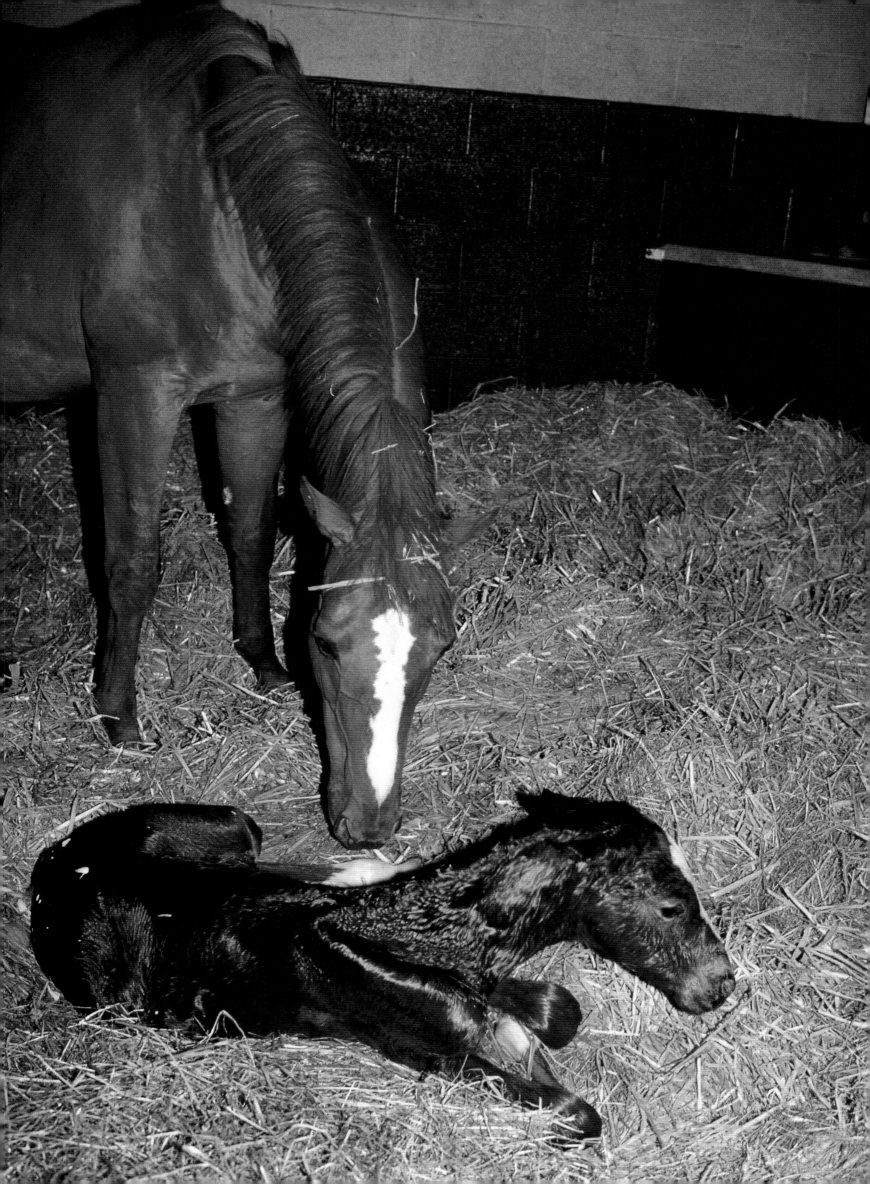

CARE AND PROTECTION

Baby animals are, by their very nature, comparatively defenseless, and are necessarily small in their early days. This not only makes them vulnerable to attack by predators, but also places them at the mercy of the elements; excessive heat or cold and heavy rain or drought can take a heavy toll on a young generation. Although some young animals, particularly those of less complex animal groups, are on their own from the outset, with many others, the role of the mother or the parents continues long after birth.

Motherly Love

Most of us are familiar with the sight of family parties of domestic pets such as cats or dogs. A mother cat with her kittens is usually a model parent and will stay with the litter except when the need to feed is great. A safe retreat will have been chosen prior to the birth and the kittens are likely to remain here for several weeks of life. They not only receive food in the form of the mother's milk, but are also kept warm and dry as well. With their poor fat reserves and compara-

tively inadequate coats, they are extremely vulnerable to heat loss. The mother also constantly grooms the young kittens. This keeps their fur clean and in good order, and also helps remove parasites such as fleas which, unless checked, might become a heavy burden for the youngsters.

Because young mammals are dependent on their mother's milk as a source of food, all mammals show some degree or other of parental care. This initial nursing stage among mammals varies according to how precocial or otherwise are the young animals in question. In deer and antelopes, it lasts but a short time; for apes and monkeys, years of dependence may be involved. As the young animals grow, however, a new stage is entered into. The baby animal may be comparatively independent when it comes to

A mother mountain goat is tolerant of her playful kid. Both mother and young have a thick coat to ward off the cold of the often hostile climate where they live.

Following page: Young ducklings such as these common or red-breasted mergansers imprint on their mother and follow her around instinctively for the first few weeks of life.

This tiny baby lowland gorilla is resting peacefully in the embrace of its mother. It will continue suckling for several months and is likely to stay with her until she has her next baby.

Three lion cubs suckle their mother's milk. The mother has to keep herself in good shape in order to be able to provide enough milk for her growing youngsters.

mobility, but still needs its mother's milk for food. Grooming and cleaning of the young animal continues as the animal grows. In addition to its functional purpose, it serves to bond the female and youngster together. This association between mother and offspring continues until the point of weaning or when the mother becomes pregnant again.

Among birds, parental care, other than feeding, is usually quite strong. For the first few days after hatching, one of the parents is likely to incubate the chicks at all times other than when they are being fed. The main purpose for this is to keep the young birds warm and dry, because at this stage, birds that remain in their nests invariably have poorly developed feathers; feathers are important to birds not only for flight but also as insulation. Incubation also serves the purpose of protecting them from attacks by predators. As the chicks get older and grow, the parents spend more and more time away from the nest gathering food, and increasingly, the chicks have to fend for themselves.

Looking on while its mother grazes on an acacia bush, a baby giraffe is still dependent upon her milk for its nutrition.

Fringed by a mane of pale hairs, a baby cheetah is confident in its mother's ability to protect it from danger. Without her, other predators such as hyenas would be likely to kill it.

Mare and foal Welsh ponies nuzzle one another. The bond between mother and offspring will last for many months.

paratively small species will fly straight at any intruder showing an interest in their offspring, aiming blows from the talons at the face and eyes. There are recorded cases of humans who have lost eyes to a defending mother owl.

An intriguing means of defense is seen among various species of plovers. This group of wading birds has representatives from around the world, and most nest on the ground. The young are well camouflaged and usually remain hidden from predatory eyes. If discovered, however, the parent birds will perform a distraction display, giving the appearance of having a broken wing. Often at great risk to themselves, they will attempt to lure the predator away from the nest site.

Young birds, defenseless as they are in their early stages, make a tasty and nutritious meal for a predator, and parent birds often go to great lengths to ensure their safety. The siting of the nest is the first means by which many birds take care of their young. If discovered by a predator, however, most parents will vigorously defend their brood by mobbing and pecking the attacker. Of all birds, owls are well known for putting up perhaps the most spirited defense of their nests. Even a com-

Moving Around

Although many species of mammal place their young in a safe den or burrow, the need will come sooner or later to move them around. Indeed, in many species, such as antelopes, this starts from a very early age. The need to transport the young also occurs in many species of monkeys and apes where the family group constantly travels within its home range. The sight of a baby baboon

Looking after a pair of cubs is a full-time job for a mother grizzly bear, whose offspring are demanding both in terms of food and attention.

Enjoying a spell on sunny weather in the brief Arctic summer, this mother and cub polar bear are relaxing together. Both have thick coats which enable them to ignore their freezing surroundings.

clinging to its mother's back is a charming one, but more extraordinary still is that of a gibbon mother swinging through the trees with a tiny infant in tow. Despite this seemingly precarious lifestyle, injuries are rare, and before very long the young gibbon can swing through the trees on its own, with all the confidence and speed of a full-grown animal.

Marsupial mammals carry their babies around in an equally pragmatic way. As previously mentioned, the baby is born at a very early stage in development, and for several months is incapable of doing anything other than remain attached to its mother's teat in her pouch. As the baby grows, it still remains in the pouch, often beyond the stage where it is completely mobile; whenever danger threatens, it returns to the safety of its mother's protection. The sight of a half-grown joey in a mother kangaroo's pouch is comic, but the strain on her muscles when she hops away from danger is severe.

With all mammals, the most intimate needs of the baby, such as feeding, are the concern of the mother. With some groups of social mammals, however, responsibility for the protection of the young falls to the family group as a whole as well as to the parents. The species concerned are herd animals with a comparatively well developed social hierarchy, such as elephants. On the plains of Africa, if danger threatens, the adult elephants bunch together and form a protective circle surrounding the young. Similar behavior can be observed among musk ox herds when the group is threatened by wolves or other predators.

From time to time, there may be occasions when even den-dwelling creatures or large carnivores need to move their young around. Such an occasion might arise if the burrow or family group is discovered by another predator. The mother fox or lion will move her cubs, carrying them one by one in her

After minutes of indecision, a young wood duck has finally decided to jump from its nest in the hole of a tree to the ground below.

When the time is right to leave the nest, the mother wood duck goes out of the nesting hole and calls to her offspring from below. These ducklings have not yet plucked up the courage to jump.

On occasion, young animals have to be forcibly moved from one location to another. Here, a mother African hunting dog is carrying a pup in her mouth.

Only a few weeks old, a timber wolf pup is enjoying the brief northern summer. It is exploring the sights and smells of a wildflower meadow which, during the winter months, will be blanketed with snow.

A baby concolor gibbon must cling on for dear life when its mother starts to swing through the trees. Despite this seemingly dangerous mode of transport, accidents are rare.

mouth, to a new and safer location. This is an important strategy since, especially as the young grow, there will be long periods when she will have to leave them alone while she goes off feeding, and it is vital that they are in safe surroundings.

Parental care is less common with groups of simpler animals, but it is certainly not unknown. Among the fishes, the mouth-brooding cichlid is a well-known example and one which is popular with aquarium enthusiasts. The male broods the eggs for protection and, once they have hatched, keeps an eye on the baby fish. If danger threatens, they retreat to the safety of his mouth. Arguably, the most charming examples of parental care are found in the sea-horses, bizarre-looking fish that live in shallow coastal seas. Male seahorses have brood pouches situated—depending on the species—either on the belly or on the tail, in which his offspring are carried around and can retreat if danger threatens.

Mention has already been made of the midwife toad, which carries its spawn around until the tadpoles hatch. In the Surinam toad of South America, this parental care is taken one stage further. The skin of adult toads is rough and pitted and the young tadpoles attach themselves to it, needing little encouragement to do so. Over a relatively short period of time, the skin grows around the tadpoles, forming protective cups in which they grow and metamorphose into tiny toadlets.

Self-Defense

Not all baby animals are entirely without the means to defend themselves. Young gazelles, for example, are capable of remarkable speed, which allows them to escape most predators except cheetahs. It could never be said that speed was one of the great assets of a young fulmar; nevertheless, it would be a foolish predator that tackled one. These dumpy seabirds nest out in the open on sea cliffs, and might be supposed to be vulnerable to attack. They are able, however, to eject the oily contents of their crops over a considerable distance, and this putrid smelling, sticky mess can easily render another bird incapable of flight.

Young owls of most species are also able to put up a spirited defense of themselves. Although their talons and beaks can inflict severe wounds, their most impressive deterrent is merely show. If a predator threatens a young owl, it will fluff up its feathers and pull its wings forward to give the appearance of an enlarged body. The sight of this seemingly huge owl, along with the loud hissing sounds it emits, is enough to discourage most would-be attackers.

Following page: Basking in the final rays of the sun, a lion cub can relax in the certain knowledge that its mother and several aunts will be keeping a watchful eye open for danger.

A mother common loon or great northern diver gives her chick a ride on her back. When danger threatens or the young bird gets chilled, this is a common response.

Often called a joey, a baby gray kangaroo still returns to its mother's pouch when danger threatens and at night. Pouches are unique to marsupial mammals.

Hitching a ride on its mother's back, a baby panther chameleon from Madagascar is gaining a degree of protection from danger.

At an early age, baby terrapins such as the red-eared slider are vulnerable to attack since their shells are not particularly tough.

A baby red-eyed treefrog makes good use of its suckered toes and its ability to jump when escaping from predators.

When danger threatens, female African elephants, including the mother, rally round to provide protection. Without this instinct, predators such as lions could be a serious threat.

Despite their huge relative size, adult African elephants seldom injure the tiny young animals. In fact, they seem to go out of their way to show consideration.

Standing in its mother's shade, a tiny baby African elephant will rely on her protection and milk for many months to come.

FEEDING AND GROWTH

From the first moments of life, the main occupation of the baby animal is feeding. This is not at all surprising, given that the youngster is small and has a relatively short time in which to grow and reach a size where it can look after itself. Developmental changes inevitably accompany the growth of the baby animal, and are equally vital to its survival. This development may involve relatively minor changes like the increase in strength and coordination afforded by improvements in the muscular system. Alternatively, profound changes may be involved, such as with amphibians, where a completely different form—the adult—metamorphoses from the tadpole stage.

Types of Food

For all the huge variation in size and form seen among mammals, they all have one thing in common: During their earliest weeks and months of life, all species feed their young on milk produced by the moth-er. This rich and nutritious fluid is produced by special glands inside the mother, and delivered through the nipples when the baby animal suckles.

Milk provides all the nutrients that a young animal needs for the initial stages of its growth and development; important ingredients include a special type of sugar called lactose, along with fats and proteins. Because different mammals have differing needs, however, the proportions of the chemical constituents varies from species to species. A quart (about 1 liter) of cow's milk, for example, contains 45 grams of sugar, 35 grams of protein, and 40 grams of fat, while in the elephant the quantities are 70 grams of sugar, 30 grams of protein, and 190 grams of fat per quart. The richest milk of all is found in the whales, with only 4 grams of sugar per quart but 95 grams of protein and 200 grams of fat.

Given the size variation among mammals, it is not surprising that not only do the constituents of milk vary from species to species, but so does the volume produced. In the largest living mammal, the blue whale, an estimated 160 gallons (600 liters) of milk is produced each day. This enables the calf, whose weight at birth is a staggering 16,000 pounds (7,250 kilograms), to

Although this young red kangaroo is probably too large to climb into its mother's pouch, it still drinks her milk. Her nipples are situated inside the pouch itself.

Like all other mammals, a baby African elephant gets its nutrition in early life from its mother's milk. Elephant milk is particularly rich.

A mother polar bear and her two cubs explore the frozen wastes of the Arctic for food. Although the cubs are still suckling, they are also able to take solid food.

surprisingly, the young animal is often reluctant to abandon this easily available source of nutrition. Among grazing animals, the transition is a gradual process, with calves of deer, cattle, and antelopes, for example, sampling the vegetation from a very early age. The young of carnivores and specialist feeders may need more encouragement or help. The young of big cats such as lions are seldom involved with the process of killing until large enough to take and deliver potentially deadly blows. They will, however, be brought to larger kills to feed or have smaller prey animals taken back to them. African hunting dogs have another means of introducing their offspring to meat: Adults swallow huge amounts of meat at a kill and, on returning to the pups in the den, regurgitate the meal.

Although many people think of baby birds as being helpless, there are many species in which the chicks are precocious. Within a few minutes of hatching they are capable of moving around on their own; game birds and wildfowl have the best known examples of precocial young. They instinctively peck at anything that even remotely resembles food, and very soon learn what is edible and what is not. Despite their apparent ability to look after themselves, however, precocial chicks are not

gain up to 175 pounds (80 kilograms) in weight per day over its seven-month suckling period.

Although milk is a vital source of food for newly born mammals, the youngster must eventually be weaned off this supply and encouraged to eat the food it will have to consume in adult life. Perhaps not

A mother and calf humpback whale swim leisurely near the surface of the sea. The comparatively limited lung capacity of the youngster means it can only make relatively shallow dives.

abandoned by their parents. They faithfully follow the mother around as she leads them to areas of good feeding, while at the same time keeping a wary eye open for danger.

The chicks of most songbirds do fit in with the perceived image of baby birds. At first, they are barely capable of doing anything other than gaping when the prospect of a parent bird with food presents itself. Insect larvae and other invertebrates are the most common food items offered to young songbirds. Even if the species concerned is a seed-eater in adult life, animal food items are likely to feature heavily in the diet of the young bird, as these are more nutritious and easier to digest.

On returning to the nest, a parent with a beak full of caterpillars distributes them to the chicks, which usually find them easy to swallow; the most demanding chicks get the lion's share of the meal. With birds of prey, a little more assistance on the part of the parent may be needed. If the prey is too large to

be swallowed whole, the parent will tear strips of flesh from the body and present them to the young. If the food is a bird, it will often be plucked prior to it being brought back to the nest. Some seabirds are similarly considerate to their offspring. Species such as Arctic terns or puffins, which feed their young whole fish, vary the size of fish they bring back according to the size of their young; larger and larger fish are brought back as the chicks grow.

Timing is also important when bringing up a brood of young birds. Most insect-eating songbirds time their nesting so that the period during which the young are being fed coincides with the period of maximum abundance of insect larvae—and, in particular, caterpillars. Not surprisingly, birds of prey such as sparrowhawks, which specialize in catching small birds, try to coincide the period when they are feeding their own young with the time when songbird broods have

A family of gray foxes suckle from their mother. With such a large family to provide for, the mother must find a constant supply of food to keep up her strength.

just fledged, and there is a relative abundance of easily caught prey. An even more interesting example of timing can be found in a bird called Eleonora's falcon, an aerobatic raptor that breeds around the Mediterranean coast. Instead of nesting in the spring like almost all other European birds, Eleonora's falcon nests in the autumn, and the period when it is feeding its young coincides with the autumn migration of the millions of songbirds which pass through the Mediterranean region.

Although it is only mammals that feed their young on milk, a few species of birds, notably pigeons, have adopted a superficially similar feeding method. Although produced in an entirely different manner, mother pigeons feed their chicks on a fluid known as "crop milk." The adult bird goes off and feeds on grain and plant material, food which would be unsuitable for the young birds. The food is digested and the nutrients absorbed; subsequently, a protein- and fat-rich "milk" is produced by the crop lining, and this is fed to the chicks in the first week or so of life. After this, they can cope with solid food and the supply of "crop milk" dries up.

In most other animals that do not show parental care, including reptiles, amphibians, and fish, the young are left to fend and feed for themselves. Generally speaking, a rule of

thumb among those that eat live prey is that if something presents itself which is smaller than you, you eat it; if it is larger, then you retreat because it may try and eat you. An intriguing alternative to this apparent free-for-all is found in discus fishes. For a period of several weeks after hatching from the eggs, the young fish stay close to the male, not only for protection but also for food; the male's skin produces copious quantities of nutritious mucous on which the young feast.

Pestering youngsters force a parent tricolor heron to regurgitate its last meal for them to eat. This means of feeding young is characteristic of many birds.

With three hungry mouths to feed, a parent great blue heron will have its work cut out for the next few weeks. Fish and frogs will comprise the bulk of the chicks' diet.

A white-tailed deer mother and fawn are feeding peacefully in a summer meadow in the eastern United States.

Following page: Eager to sample its mother's diet, a baby orangutan tries to snatch a morsel of food from her grasp, but milk will still be its primary source of nutrition for some time to come.

Growth and Development

Although the young of some mammal species are, of necessity, able to take care of themselves within a short time of being born, most are extremely vulnerable. Mice, for example, are almost naked when born, as well as being blind; without food and warmth from the mother, they have no hope of survival. Under favorable circumstances, however, they put on weight rapidly, develop a thick and warm coat of fur, and open their eyes. Within a matter of weeks they are fully grown and independent, and are also sexually mature before very long.

With larger mammals, the period of association with the mother or both parents can be much longer. Suckling can last for many months, and the period between weaning and independence can extend from a few months to several years, depending on the species. True independence is often not reached until the onset of sexual maturity, and in primates (including man) may not occur until well after this time.

Among birds, precocial chicks are born with their eyes open in both the literal and metaphorical senses. Their growth and development progresses rapidly and is entirely dependent upon their own ability to feed. In most species, within a few weeks the downy, fluffy feathers so characteristic of chicks will have been replaced by proper feathers, which both provide better insulation and the power of flight. Once the power of flight has been achieved, the young often become independent, although family parties of game birds,

Bobcat kittens make a charming sight as they seemingly enjoy each other's company. This behavior is, however, likely to be a prelude to a lively scrap.

for example, may stay together well beyond the breeding season. The most striking examples of persisting family bonds among birds with precocial young are seen among many species of geese and swans. Whooper swans, for example, which breed in the high Arctic, migrate thousands of miles south in autumn in family groups which stay together throughout the winter months.

Chicks that hatch naked and with their eyes closed are referred to as altrical; they are entirely dependent upon their parents for both food and warmth. The response of such a chick to any form of disturbance that could herald the arrival of a parent is to open its beak and beg for food. This behavior continues until the point where the young have acquired a full set of feathers and can fly. Many such birds are fed by their parents even after they leave the nest, although for others this important event severs the parental ties.

The length of time it takes for a newly hatched chick to become fully grown and independent varies from species to species, although it is true to say that, roughly speaking, the larger the bird the longer this process takes. A small warbler, for example, may take three weeks to be independent whereas it may be as long as nine months before a waved albatross chick flies away from its nest site.

Although lively and active for much of the day, most lynx kittens tire easily and spend long periods sleeping or resting.

An exuberant leap may seem simply like fun but for baby cheetahs, such activities are a vital part of training for the world at large.

The rough and tumble of play teaches baby cheetahs skills they will need in later life when tackling prey for the first time.

The wide-eyed apparent innocence of a baby cheetah belies the fact that this is a killer in the making. Hunting skills are learned as the cub plays with its peers and watches its parents.

This graceful treefrog from Australia is changing from a tadpole into a miniature adult. Within a few days it will have absorbed and lost its tail.

A baby pygmy chimpanzee or bonobo takes its first faltering steps on two legs. It will not be long until it returns to the care and safety of its mother's embrace.

Going up may be easy, but this young grizzly bear is about to find out that climbing down is not so straight-forward. Experience will soon teach it its own limitations.

DEVELOPING LIFE SKILLS

Because of their small size and inexperience, baby animals are by their very nature vulnerable in the world at large. Mammals and birds rely, at least to some extent, on their parents for a supply of food and some degree of protection from attack in their early days. Sooner or later, however, the young animal will have to start fending for itself, and its chances of survival will depend heavily on its ability to find its own food and respond to threats of danger. The ability to feed is not only of direct importance for the baby animal's immediate survival, but can also have future consequences. Among mammals which hibernate through the winter, there is a critical weight which must be achieved in the autumn in order for their fat reserves to last through the winter. Obviously, this critical weight varies from species to species, but if the baby animal does not achieve it, it will not survive. Not surprisingly, therefore, the development of life skills features prominently in the daily lives of baby animals.

Innate Skills

Although many life skills have to be learned over a period of time, some are part of the innate behavioral patterns of the baby animal from birth or from the point when they are needed. In baby mammals the ability to suckle needs no instruction and starts from a very early age. Beyond the milk-drinking stage, specialist feeders often need to develop their feeding skills, but more generalist feeders such as grazing animals start nibbling at vegetation early on.

Among baby birds, the ability to feed is innate among species which leave the nest immediately, as is the reaction to gape and beg for food among chicks that remain in the nest. Although obviously improving with experience, the ability to fly is also inborn. For some birds, a few initial mistakes will not result in any harm, but for others, such as those that nest on high cliff ledges, an inability to fly on a first attempt is likely to be fatal.

Another extraordinary innate ability found in birds is that of navigation, a skill which is

Although it may look rather forlorn, a baby tree porcupine from Amazonian Peru is an adept climber, with powerful feet and claws.

Alert and curious about the world around it, a lynx kitten is exploring its surroundings. Undertaken under the watchful eye of a parent, this activity is a vital way for young animals to learn about their environment.

particularly important among species that migrate soon after fledging. Many songbirds migrate at night and the guidance of adult birds is clearly unlikely to occur. Even more extraordinary is migration in the European cuckoo. Adult cuckoos lay their eggs in the nests of other birds, and will have embarked on their migration back to Africa long before their offspring are even ready to fly.

Nevertheless, most of the juvenile birds still make the same journey successfully.

Many baby animals also show innate responses to danger. From a very early age, the instinctive response of a baby hedgehog is to roll itself into a tight ball, protected by its spiny coat; it needs no instruction in this. A more common response to danger among baby mammals is to return to the

A baby spider monkey looks somewhat apprehensive at having to cling to its mother's back. Its firm grip will not be relinquished until she settles to feed or rest.

With its mother off foraging for food, a baby raccoon is eyeing its surroundings. It will not be long before it starts to venture out with its mother.

mother or perhaps to the den or burrow; this may mean one and the same thing for many mammals. The response may be initiated by the threat of danger itself or by an alarm call from an adult.

Sound also plays an important role among baby birds. For example, among species where the young remain in the nest, a common response of the chicks to a slight disturbance at the nest is to gape and beg noisily for food. This same response can usually be quelled by alarm calls from the parents if a predator is around—the young birds remain motionless until the danger has passed. Parental alarm calls have a similar effect among many ground-nesting birds such as stone curlews or ringed plovers; the sound causes the chicks to lie flat and motionless on the ground.

Imitation and Play

Among humans, many of the skills we develop as children—and even as adults—are learned by watching others and imitating them. It should come as no surprise, therefore, that the same thing happens in many other animals as well.

Among monkeys and apes, imitation of adult behavior is particularly obvious, and can at times be almost comical. Species such as vervet monkeys or baboons which have a varied, omnivorous diet often have to learn special techniques for obtaining food. The knowledge of which fruits are good to eat and which are unrewarding is acquired more easily by following the examples of others in a group. Also, learning the way in which a particular tough-skinned fruit can be

Following the example of its mother, a baby Florida sandhill crane is looking for food. Parental imitation is an important means by which many young creatures learn to feed.

Two Japanese macaques are engaged in rough play. Fortunately, their thick coats will keep them warm if one of them should fall into the snow below.

Its mother's tail is too inviting a plaything to avoid for this lion cub. Other adults in the pride may not be so tolerant of this intrusion into personal space.

A roar of disapproval should be enough to tell this lion cub that its play may have gone too far. Male lions are not nearly so tolerant as females of youthful exuberance.

opened is often done as much through imitation as by trial and error.

Among chimpanzees, some feeding techniques are highly specialized and are not used by all animals, and particular groups pick up techniques developed by individual animals. The use of flat base and hammer stones to crack open hard-shelled nuts, for example, is not an innate skill—not all chimpanzee groups use this method. In those groups that do, however, offspring often learn the technique by watching their mothers and by using precisely the same stones. Even more fascinating is to watch is the way in which chimps "fish" for termites using a fashioned grass stem inserted into a termite mound; the angry termites bite onto the intruding stem and are withdrawn and eaten by the chimp. Both the fashioning of this tool and the method itself have to be learned by each new generation of chimps, and they do so by imitating their parents.

Every human mother appreciates the need for children to play, and the fact that the pastime is as much a learning process as a means of letting off steam. So it is with many other baby mammals. Among domestic cats and dogs, kittens and puppies all engage in mock fights and stalk one another. Precisely the same thing happens among their wild cousins, and lion cubs are forever getting into scraps or chasing one another's tails. With lions and all other wild animals, this apparent play is a vital way in which they begin to learn skills more useful in later life. The ability to stalk prey, to suffocate by strangulation, or to disembowel with blows from the hind legs can all be observed on a playful level among young cats. For some mammals which live in large social groups and in which a definite hierarchy exists, play among youngsters also serves to establish a pecking order that persists into adult life. For these animals, the appearance of play as youthful exuberance is only surface deep; the underlying reasons are profoundly significant.

Play is a vital part of a young animal's upbringing. Two baby African elephants are enjoying each other's company while their parents feed.

Two lion cubs, photographed in the Masai Mara in Kenya, are beginning to explore the world around them. Already at this early age they are skilled and fearless climbers.

Parental Instruction

Although young animals are observant and acquire a number of life skills by simple observation and imitation, in some species the parents actively educate their offspring in the ways of the world.

One of the most striking examples of this is found in cheetahs. When a mother cheetah has cubs which are about six or seven weeks old, she may bring small prey, such as a young gazelle, back alive. She will then leave the hapless animal for the cubs to practice their hunting skills on, retrieving the gazelle should it escape from the inexperienced cubs. They will practice stalking the victim, felling it with their front paws and claws, and killing it by strangulation. Although this behavior may seem gratuitous to our eyes, it is essential that the cubs learn how to kill for their future survival.

Other predators learn from their mother's example how to catch and kill their prey. Mother raccoons, for example, show their offspring by example how to catch difficult prey such as crayfish. The same procedure

The simplest way for this mountain lion to get an unruly youngster from one place to another is to carry it in her mouth.

Piggyback rides on its mother's back are a comfortable and convenient way for this low-land gorilla to get around. Once it gets much bigger, the mother will be less inclined to carry its weight.

Getting in training for a laid-back lifestyle, a young orangutan is enjoying a relaxing rest on its mother's back.

A mother Alaskan brown bear has brought her two cubs to a salmon river in order to show them the skills of fishing. Even for a bear, this is a process that takes time and experience to perfect.

A young humpback whale will remain with its mother for up to a year, and it will follow her from its calving grounds in Hawaii all the way to Arctic waters.

Visiting the water for the first time, two grizzly bear cubs are nervously watching their mother fish from the safety of the river bank.

can be seen in grizzly bears which congregate beside certain Alaskan rivers in summer to catch migrating salmon. Not only do the bear cubs learn by example how to catch such slippery prey, but they are also shown the best spots to fish by their mothers.

Independence

There comes a time in the lives of all young animals when they are no longer babies and must fend for themselves in the world at large. For some more primitive creatures, this happens early in life, perhaps at the egg-laying stage or soon after hatching. Among birds, the break usually occurs soon after fledging (although some family groups stay together), but in mammals, the bond between mother and young continues for much longer. Even here, however, the ties must be severed eventually in most species.

Some mammals, such as most monkey and ape species, live in social groups. When the ties between mother and young are finally broken, this may not necessarily mean that the pair part forever. Females may spend the rest of their lives living as part of the same group as their mothers, even raising their own offspring side by side. For young males, however, it may be rather different, because they are seldom tolerated by the dominant male as they approach sexual maturity. They may well be forced to leave the group in which they were raised and fend for themselves, either in a solitary manner or, more usually, as part of a roving group of similarly aged young males.

For solitary mammals, and predatory species in particular, a territory is often defended whether the animal is male or female. Not surprisingly, there comes a time when even the ties felt by the mother are not enough to maintain the bond between her and her young. Competition for food in particular, as well as safe retreats within the territory, force her to drive the young animal away. Now comes the first big test for the youngster: No longer a baby, it must take its place as an adult or perish in the process.

Its mother's thick coat of fur is ideal for a baby collared lemur to cling to. The two will stay together for many months, until the youngster is able to fend for itself.

It is not only mother olive baboons that take an interest in their babies. Other members of the group also show curiosity and concern for this youngster's well-being.

INDEX

*Page numbers in **bold-face** type indicate photo captions.*